Frank Rutley

**The Felsitic Lavas of England and Wales**

With an Introductory Description of the Chief Characters of This Group of Rocks

Frank Rutley

**The Felsitic Lavas of England and Wales**
*With an Introductory Description of the Chief Characters of This Group of Rocks*

ISBN/EAN: 9783743418318

Manufactured in Europe, USA, Canada, Australia, Japa

Cover: Foto ©berggeist007 / pixelio.de

Manufactured and distributed by brebook publishing software
(www.brebook.com)

Frank Rutley

**The Felsitic Lavas of England and Wales**

# MEMOIRS OF THE GEOLOGICAL SURVEY.

## ENGLAND AND WALES.

# THE FELSITIC LAVAS

OF

## ENGLAND AND WALES,

WITH AN

## INTRODUCTORY DESCRIPTION OF THE CHIEF CHARACTERS OF THIS GROUP OF ROCKS.

BY

### FRANK RUTLEY, F.G.S.,

LECTURER ON MINERALOGY, ROYAL SCHOOL OF MINES.

PUBLISHED BY ORDER OF THE LORDS COMMISSIONERS OF HER MAJESTY'S TREASURY.

## LONDON:

PRINTED FOR HER MAJESTY'S STATIONERY OFFICE;

AND SOLD BY

LONGMANS & Co., Paternoster Row; TRÜBNER & Co., Ludgate Hill;
LETTS, SON, & Co., Limited, 33, King William Street, E.C.;
E. STANFORD, 55, Charing Cross; WYLD, 12, Charing Cross;
and B. QUARITCH, 15, Piccadilly.
Manchester : T. J. DAY, 53, Market Street;
Edinburgh : W. and A. K. JOHNSTON, 4, St. Andrew Square;
Dublin : HODGES, FIGGIS, & Co., 104, Grafton Street; and
A. THOM & Co., Limited, Abbey Street.

1885.

*Price Ninepence.*

# CONTENTS.

u  18642.    Wt. 8917.                    A 2

# PREFACE.

In the following pages a summary is given of all that is at present known regarding the distribution of that interesting group of rocks, the Felsitic Lavas, in England and Wales. Mr. Rutley has incorporated the published observations of other petrographers with those made by himself in the Collections of the Geological Survey. As the number of instances of devitrified rocks is constantly being increased in the British Islands and abroad, and as the subject of devitrification is one on which much light still requires to be thrown, both by observations on ancient rocks and by laboratory experiments, it is believed that the publication of this Memoir will be of service.

ARCH. GEIKIE,
Director-General.

Geological Survey Office,
28, Jermyn Street, S.W.,
21st November 1885.

# THE FELSITIC LAVAS

OF

## ENGLAND AND WALES,

WITH AN

## INTRODUCTORY DESCRIPTION OF THE CHIEF CHARACTERS OF THIS GROUP OF ROCKS.

### INTRODUCTION.

Ten years ago the former occurrence of vitreous rocks in England and Wales was not demonstrated, even if suspected. The terms felstone, felsite, petrosilex, hornstone, eurite, &c., had for many years been applied to rocks of more or less compact texture, often with conchoidal fracture, having approximately the hardness and fusibility of felspars, and believed to be intimate admixtures of felspar and quartz. Such rocks frequently exhibited, especially upon weathered surfaces, a finely banded structure, closely resembling the lamination of sedimentary rocks, and where, as sometimes happened, they were found to be interbedded with ancient stratified deposits, occasionally fossiliferous, it is not surprising that they were now and then described as " altered flags " and slates, to which, when in advanced stages of metamorphism, these felsitic rocks certainly bear a somewhat close resemblance. Their mineralogical affinity to the groundmass of quartz-porphyries or quartz-felsites and also to trachytes was recognised. Many of them were mapped by the Geological Survey as lavas, and more than twenty years ago, the banding in one of the Welsh felsites was stated by Sir Andrew Ramsay to have " probably originated in the same cause that produced the lamination in the lava of Ascension."

Although a very large proportion of the English and Welsh felsites are now known to be true lavas, it should also be remem· bered that many rocks of quite different origin exist which, from their mineral constitution and structure, may also be termed felsites or felstones. To describe all the rocks which might be included in the group of felstones would be a somewhat difficult and arduous

task, even if the needful information were at hand ; although it
may, perhaps, be useful to give, in conclusion, a few selected
examples of felstones which have apparently had an origin
different from that of the lavas which specially constitute the
subject of these pages.*

It has already been stated that felsites are believed to consist of
felspar and quartz. The constituent grains of these minerals are
very minute in rocks of this class, but when we examine thin
sections of felsite under the microscope, polarised light being
employed, we notice that there is a considerable difference in the
size of the crystalline grains in different kinds of felsite ; we also
see that there is sometimes a marked difference in the distinctness
with which the boundaries of the grains are visible, for while in
some cases their outlines are quite sharply defined, so that the eye
can clearly recognise the exact area occupied by each grain in the
section, as in the groundmass of an ordinary quartz-felsite or
elvan, in others the grains seem to shade off one into another, so
that no distinct boundary can be assigned to any particular grain,
in which case the section presents a hazy or blotchy appearance,
while, furthermore, we may in some instances meet with so fine a
texture that the field of the microscope seems merely to be
studded with minute points of light even under moderately high
powers. It is evident, then, that we may have very different
" polarisation pictures " resulting from different varieties of felsite.

The first of the structures just mentioned, in which the constituent grains are distinctly marked and the several quartz and
felspar grains are capable of recognition under the microscope,
is termed a " micro-crystalline structure."

The second and third, in which the boundaries of the crystalline
grains are inappreciable and the different mineral characters of the
grains cannot be determined, is known as a " crypto-crystalline
" structure."

Both the quartz and the felspar exhibit double refraction, but
between these doubly-refracting or anisotropic grains a substance
may sometimes be detected in felsites which does not exhibit any
double refraction ; it is inert upon the polarised beam ; it is
isotropic. Such isotropic matter may be perfectly homogeneous.
This is glass or the " glassy basis " of Rosenbusch.† In its truly
vitreous condition it remains absolutely dark during a complete
revolution of the stage between crossed nicols.

In other instances isotropic matter is present which is not
homogeneous and structureless, but seems to consist of diminutive
fibres, scales, and granules with some interstitial substance. This
devitrified substance is the " micro-felsitic basis " of Rosenbusch.

---

* It may be of interest to mention that a straight line drawn from Dunstanborough
Castle, N.E. of Alnwick to Skomer Island, off the coast of Pembrokeshire, passes
through some of the principal English and Welsh localities where felsitic lavas of
Silurian age are met with.
† Mikroskopische Physiographie der Massigen Gesteine, p. 71. Stuttgart, 1877.

Any of the conditions above mentioned may occur in felsites, constituting what is understood by the general term "felsitic matter."

Felsites, however, frequently exhibit other structures, some of which are very characteristic and indicative of the conditions under which the rocks were formed.

*Banded Structure.* This structure is often met with in felsites of different kinds and sometimes of different origin. It is frequently seen in those felsites which were once vitreous, and in almost all such cases it may safely be inferred that the banding existed prior to devitrification. In unaltered vitreous lavas, erupted within comparatively recent times, the banding may be of various kinds. Its simplest form consists, perhaps, in slight differences in the character of the glass itself, evinced only by difference in colour. In the obsidian of Ascension, for instance, when a thin section is examined by ordinary transmitted light, dark brown bands may sometimes be seen traversing a pale yellowish or colourless glass. Frequently these bands have no sharply defined margins, but fade gradually into the adjacent glass as though the latter had been merely dyed with some brownish pigment. At other times such bands of coloured glass may have very distinct boundaries, as in an obsidian from the Yellowstone District, in which sharply defined bands of deep yellowish or orange coloured glass traverse a glass of very much paler tint.

Such glass bands sometimes depolarise, owing to strain, and between crossed nicols they extinguish in the direction of the fibres, *i.e.*, when the fibres are brought parallel to the principal section of one of the nicols the light, which was previously transmitted by the fibres when in other azimuths, now becomes extinct. It is a common thing to find such bands densely crowded with the small crystalline bodies termed microliths or crystallites, some of which may be seen to exhibit double refraction while others show none; also with minute spherical, isotropic bodies termed globulites.

Frequently, too, in unaltered glassy lavas, distinct bands are visible which, when examined under high powers, are seen to consist almost exclusively of streams of microliths, without any appreciable difference between the glass of the band in which they occur and that adjacent to it, the glass being colourless, or its tint uniform throughout (Fig. 7, Pl. 1.) Such microlithic streams or bands may be seen in some of the perlitic lavas of Saxony.

It would seem unreasonable to expect to recognise banding, of the kinds just described, in devitrified rocks, since one would assume that the superinduced crystalline structure would obliterate all evidence of banding shown by mere difference of tint in the glass, unless there were some sensible difference in the chemical composition of the more or less deeply coloured, or of the coloured and colourless bands. And, so far as the bands of microliths are

concerned, we can well imagine that, although, in common with
larger crystals, they would survive the changes brought about by
devitrification, yet we can hardly expect to recognise the in-
dividuality of these diminutive bodies when surrounded by and
intermixed with the products of a subsequent crystallisation. It
is quite possible, however, that differences in the state of tension
of the material in and about the microlithic bands may give rise
to marked diversity in the texture of the crystalline products of
devitrification. Great differences in crystalline texture may often
be noted between the bands of devitrified lavas and the parts
separating those bands, the latter, as a rule, exhibiting a much
less coarse grain than the bands themselves (Fig. 8, Pl. 1).
Indeed, some difference in grain is always to be found where this
laminated structure exists.

The bands may be either approximately parallel or sinuous, and
it is common to find that the presence of a porphyritic crystal
or a large spherule is accompanied by more or less deflection of
the adjacent banding.

This phenomenon, taken in conjunction with the axially parallel
arrangement of microliths in the bands, has given rise to the terms
"fluidal structure," "fluxion structure," or "flowage" as it is
sometimes termed; the streams of microliths and other bands
being assumed to sweep past and around these larger bodies as
the water of a river sweeps round an island (Fig. 7, Pl. 1).
When structure of this kind is present in a felsite it is usual to
accept it as evidence, partially confirmatory, of former vitreous
character, but it should be borne in mind that similar phenomena
are met with not only in trachytes and other crystalline lavas,
but also in some dykes, and even in metamorphic rocks, as pointed
out most forcibly in the recent works of Professor A. Geikie* and
and Dr. J. Lehmann†; while like appearances may be noted
in boulder-clays and other stratified deposits where large un-
yielding masses cause a synclinal deflection of the subjacent, and
an anticlinal one of the incumbent beds.‡

Sometimes the banding in felsitic rocks, and also in their more
modern glassy representatives, becomes not merely sinuous, but
contorted, and, in rare instances, contortion may reach even the
pitch of intricate convolution, in which case a "damascened"§
structure results (Fig. 10, Pl. 1).

Bands composed of spherules are often met with both in glassy
and in devitrified lavas. In the latter case the spherules are
frequently the product of devitrification, and it is not unusual in
a devitrified, banded lava to find that the spherules which devitrify

* "Text Book of Geology," 2nd Edition, see 'Shearing of rocks,' pp. 289, 506,
531, &c. Fig. 246 in this volume shows the close resemblance in the structure of
a compressed Cambrian sandstone to that of a rhyolite.
† "Untersuchungen über die Entstehung der Altkrystall, Schiefergesteine."
‡ In a paper by the late Rev. J. Clifton Ward (Quart. Journ. Geol. Soc.,
Vol. XXXI., p. 410), an "altered ash" is described, in which a structure, sug-
gestive of fluxion, is attributed to an alteration of the rock.
§ "The Study of Rocks," 2nd Edition, p. 181.

the bands differ considerably in size from those which devitrify the rest of the rock. Occasionally they are much smaller in the bands than elsewhere.

The banded structure of felsites is frequently rendered evident by superficial inequalities formed upon weathered surfaces, the rock being more or less deeply eroded in fine and approximately parallel furrows. Since similar banding and erosion may be met with in altered sedimentary rocks, which, when in the condition of porcellanite, offer just as good conchoidal fractures as felsites, it is dangerous to base any positive conclusions upon this class of evidence.

*Spherulitic Structure.* The development of spherules constitutes sometimes the first, at others the last change in a glassy rock. In the first case it represents a structure preceding, or coeval with the solidification of the lava, and it may be restricted to the development of isolated spherules or of single belts of spherules, which may be separate or may have coalesced in such a manner that their sections appear as beaded strings (Fig. 11, Pl. 1); or the union may have been so complete that no constrictions indicative of beads are visible, in which case they form a band with straight and parallel edges (Fig. 12, Pl. 1). Spherules may also become clustered together in great numbers, forming bands many spherules thick (Fig. 9, Pl. 1), and occasionally they group themselves around porphyritic crystals and along cracks in the rock.* Sometimes the development of spherules is so great at the period of solidification of a lava that they constitute the entire rock (" *Sphärolithfels* " of German petrographers). In other cases a vitreous rock, which may possibly have been quite free from spherules when it solidified, may have become completely devitrified through the formation of these bodies, as in the devitrified obsidian of Till's Hole (described on p. 12).

The spherulitic structure of a rock is sometimes rendered evident through weathering, the weathered surfaces often looking as if they were studded with whitened peas or shot, and occasionally the spherules are so large that they may be likened to tennis balls. It would be useless to attempt to classify these bodies according to their size, for they vary from two or three inches or more in diameter to absolutely microscopic dimensions. In the case of spheroidal concretions or separations, it may be a matter of feet rather than inches.

In a paper on artificial devitrification by Mr. D. Herman and the author†, instances are given in which pieces of plate glass, submitted to a high temperature for a few hours only, have had merely the surfaces slightly affected. Such surfaces show that exceedingly minute spherical bodies ("globulites" of Vogelsang) have been developed in great numbers, and, when the film in which they occur is examined microscopically under high powers,

---

* Quart. Journ. Geol. Soc., Vol. XL., p. 344.
† Proc. Royal Soc., Vol. XXXIX., pp. 87–107. Plates 1–4.

they are seen to have massed themselves together in certain spots, giving rise to little granular segregations (Fig. 6, Pl. 1). From an examination of many specimens of glass, operated upon for varying periods in a similar manner, some of which contain well-developed spherules with a radiating crystalline structure, it seems more than probable that the little groups of globulites represent an embryonic stage in the formation of spherules. Vogelsang[*] states that he has seen a crystal increase by the accretion and transformation of globulites, and he has noticed, in the course of his experiments, that globulites, so long as they remain isolated in the magma in which they have been formed, remain isotropic, but that, as soon as they become grouped together into crystallites, double refraction usually becomes apparent and that this property is only feebly developed when the crystallites are imperfect. There is, then, under such circumstances, a passage from the amorphous to the crystalline condition, but when, through increased rigidity, the resistance of the solidifying magma becomes augmented, the globulites, even when in actual contact with a crystal, cease to be assimilated by it, and permanently retain their spherical form.

Globulites are sometimes seen to be densely grouped about spherulites (as in Fig. 2. Pl. 1) so as either to form a more or less thick envelope around them, or in great part to constitute the spherulite. Examples may be seen in the obsidians of Vulcano and the Rocche Rosse, and it is interesting to note that this globulitic matter depolarises in a manner indicative of a radial crystallisation. A careful examination of many of these spherules shows that there is, indeed, a divergent crystallisation set up within these globulitic masses. When globulites come into contact and arrange themselves in lines, as they frequently do, such linear aggregates have been called "margarites," from their resemblance to strings of minute pearls; while, when the globulites in such linear order become merged into one another, so that the globular forms disappear, and minute, elongated, cylindrical bodies result, the latter are termed "longulites."

It would seem, then, that in the cases just cited we have longulites present, mixed with globulites which may or may not have a linear arrangement, and it is probably to the presence of divergent groups of longulites that the double refraction is due; since, although it might be difficult to prove the existence of this property in a single longulite, yet it is possible that bundles of them may exert an appreciable depolarising effect. The same may be said of the depolarisation consequent upon strain, which is visible in a bundle of fibres of spun glass, while in a single fibre no depolarisation is apparent. If the globulites be solid microscopic pellets of vitreous matter,

---

[*] "Die Krystalliten." Bonn, 1875, and "Archives Neerlandaises," 1872; also Zirkel's Mik. Beschaff. der Min. u. Gest, p. 95.

as assumed by Vogelsang and others (Link[*] and Weiss[†] think otherwise), then we must regard the longulites as merely cylindrical glass rods or spiculæ, in which case the comparison of a fibre of spun glass with a longulite becomes admissible. If, however, we regard globulites merely as pellets of vitreous matter it is difficult to see how glass can become devitrified through their development, and, that these minute bodies are endowed with properties which mere glass pellets do not possess, seems probable from their methods of grouping, and from the double-refraction resulting from their coalescence.

Spherulitic structures have been induced in glass, not merely through the agency of dry heat, but also by heating under considerable pressure and in presence of water. Such experiments made by Prof. Daubrée,[‡] approximate, perhaps, more closely to natural conditions than the dry heat process,[§] but the results in both cases are interesting since they show that similar structures may be produced under somewhat different conditions.

Spherulites, as a rule, show no distinct nucleus, the crystallisation appearing to have started from a point, but in some cases a crystal or a group of crystals may be seen to form the nucleus. Slight differences in the chemical composition of artificial glass have been found by Mr. Herman to influence the production of spherules, some kinds devitrifying throughout without the formation of a single spherule, while in others of only slightly different composition spherules are developed in profusion.[||]

The globulitic portions of some of the spherules in the obsidians from the Liparis, already mentioned, are of a deep brown colour when viewed in thin section by ordinary transmitted light. If this colour be inherent in the globulites themselves it would indicate that either there is some difference chemically, or in degree of saturation, between these bodies and the surrounding glass, or that the molecular condition of the former is more dense than that of the latter, in which case a greater number of molecules would occur in the globulitic portions of the section than in the glass which is comparatively free from them, the result being equivalent to an augmentation of tint produced by the superposition of similarly coloured plates. Rocks are sometimes partially devitrified by the development of globulites.[¶]

Spherulites are occasionally seen to be surrounded by glassy borders[**] (Fig. 1, Pl. 1). They more frequently, however, have no such borders, and possess merely a radiating crystalline structure (Fig. 3a, Pl. 1), which is best, and often only, revealed

---

[*] Pogg. Annal. XLVI. Ueber die erste Entstehung der Krystalle, 1839.
[†] Ib. CXLII., p. 324.
[‡] Etudes Synthétiques de Géologie Expérimentale, p. 169, et seq.
[§] Quart. Journ. Geol. Soc., Vol. XLI., 1885; Presidential Address by Prof. Bonney.
[||] Herman & Rutley, Proc. Royal Soc., Vol. XXXIX, 1885.
[¶] Zirkel, Mik. Beschaff. d. Min. u. Gest, p. 274.
[**] Rutley, Monthly Microscopical Journ., April 1876.

in polarised light, and, like other radiating crystalline aggregates, they show a dark interference-cross when viewed between crossed nicols (Figs. 3b, and 9, Pl. 1). This is the most common type of spherulite.

Less common, and usually of considerably larger size, are those which show no indication of a radiate structure, but appear under tolerably high powers to be made up of minute granules (Fig. 5, Pl. 1).

Sometimes spherules exhibit irregular, or denticulate, or serrated boundaries (Fig. 4, Pl. 1), as in some of the devitrified lavas of the Glyders in North Wales. These have always a radiating crystalline structure, and it appears as though their peripheral development had been influenced irregularly by the crystallisation of the surrounding matter.[*]

Occasionally elongated spherules are met with, which Zirkel has . termed axiolites, (Fig. 13, Pl. 1). Axiolitic structure has not hitherto been seen in British rocks. The largest spherulitic bodies often present quite a nodular character. These are the "lithophysen" of Von Richthofen,[†] who considered them to have resulted from the inflation of siliceous segregations by imprisoned steam. They are frequently cavernous.

Professor Bonney, who, among others, has investigated the nature of these bodies, has arrived at the following conclusions:—[‡] " That the nodular or spheroidal structure has been produced in two ways."

> a. " By simple contraction and roughly concentric cracking of the mass in cooling, being thus intermediate between the perlitic structure common in glassy acid lavas and the spheroidal structure common in basalt, which, so far as I know, is rather rarer in the former."
>
> b. " By similar contraction in cooling, which is determined by the presence of a cavity, and produced as follows:— When the cavity is first formed we may regard the whole viscid mass in the neighbourhood as in a state of equilibrium between the various forces acting on the cooling lava (contraction, &c.), and the pressure of the gaseous contents of the cavity. As cooling proceeds (uniformly suppose) the volume of the latter diminishes rapidly, and its pressure against the walls of the cavity decreases. The various forces are no longer in equilibrium, and the contractile strain will be relieved by the formation of a crack, roughly concentric with the cavity which, as we might expect, is more regular than it in form."

" That the cavities are then filled, wholly or partially, by infiltrated minerals in the usual way."

---

* Quart. Journ. Geol. Soc., Vol. XXXV., p. 508.
† " Jabrb. k. k., Geol. Reichsanst." 1860.
‡ Quart. Journ. Geol. Soc., Vol. XXXVIII., p. 289.

"That the nodules, thus rendered more solid (and in other cases from the effect of their form, aided perhaps by extremely minute differences in texture, due to the disturbance of equilibrium in cooling), produce the usual distortion of the cleavage planes when the whole mass is compressed."

Some observers believe these bodies to be merely large spherulites. Mr. Cole, in a recent paper,* adopts this latter view, regarding them, at all events in some cases, as the result of devitrification which took place during cooling, and he adduces among other reasons, the presence of a radial structure and the occurrence of smaller spherules within these large ones. He does not consider that they have been formed subsequently to the consolidation of the rock in which they occur. Analyses made by K. von Hauer show that there is little difference between the chemical compositions of these bodies and that of the matrix in which they occur. Specimens from the Glyders, Conway, and other localities in North Wales are deposited in the Rock Collection of the Museum of Practical Geology. Figures of the microscopic structure of spherulitic lavas will be found in the following volumes of the Quarterly Journal of the Geological Society, Vol. XXXVII., Pl. XX., XXI.; Vol. XXXIII., Pl. XX.; Vol. XXXVIII., Plate X.; Vol. XL., Pl. XVIII.; and Vol. XLI., Pl. IV.

*Perlitic Structure.* The delicate spheroidal cracks often occurring in vitreous and in devitrified lavas constitute what is known as perlitic structure. These cracks are seldom or never quite continuous so as to form perfect spheroids; they are approximately concentric, but overlap one another so as to produce a series of somewhat irregularly concentric shells, rather like the layers of an onion. The rocks in which this structure occurs are usually traversed by minute reticulating, rectilinear fissures, between which the spheroidal, perlitic bodies lie (Fig. 14, Pl. 1). These bodies occasionally show depolarisation, due to strain, but this phenomenon is only evident in moderately thick sections. It serves, however, to confirm what was previously surmised, namely, that perlitic structure is the result of contraction.† The perlitic fissures are quite independent of other structures in the rocks in which they occur, and are commonly seen to traverse streams of microliths, &c., constituting the so-called fluxion structure, and they were probably formed at the close of, or subsequently to, the period of solidification of the vitreous rock. It is a structure not peculiar to the highly silicated vitreous rocks, but also occurs in tachylyte, and is met with in various amorphous or colloid substances, as in Canada balsam, varnish, &c. Although the perlitic fissures are often extremely delicate, devitrification seems to fail, as a rule, to

* Quart. Journ. Geol. Soc., Vol. XLI., p. 163.
† Quart. Journ. Geol. Soc., Vol. XXXII., p. 140., Vol. XXXIII., p. 452, and Monthly Microscop. Journ., Vol. XV., p. 180.

obliterate them, the perlitic structure in some of the oldest-known
lavas being as perfectly preserved, in spite of devitrification, as in
glassy volcanic rocks erupted in comparatively recent times.

---

Having now briefly reviewed the leading structural characters
of felsitic lavas and their modern representatives, we may proceed
to the description of some of the principal examples which have
hitherto been met with in England and Wales. The geologist
cannot hope to restore clearly the physical geography of the far
distant ages when these felsites were molten streams, but it has
long been known that lava-flows of considerable extent, and
sometimes of considerable thickness, were poured out in very early
times, and it has since been ascertained that those of the class
which forms the subject of this Memoir, cooled down into
rocks often identical with the obsidians, perlites, and pitchstones of
much later date, as was first indicated, from microscopical observa-
tion by Mr. S. Allport[*] in 1877.

The figures on Pl. I. are not drawn from nature. They are
diagrammatic, but sufficiently like the structures they are intended to represent, and will probably serve their purpose better
than actual drawings from microscopic preparations would have
done.

The figures in the other plates are given in little more than
outline, in order to suit the method of reproduction employed.

The microscopic textures and small spherulitic structures are
not as a rule appreciable by ordinary transmitted light, while the
method of shading adopted being quite inadequate to express the
appearance of most of these structures in polarised light, it has
been thought better to refrain from any attempt to represent them
except in the diagrammatic sketches on Pl. I.

References are, however, given in the text to works in which
these phenomena are properly figured, and in addition to these,
admirable drawings will be found in Zirkel's "Microscopical
Petrography of the 40th Parallel," and Fouqué and Lévy's
"Minéralogie Micrographique."

In the following pages the term "perlite" is not employed as a
rock-name, since perlitic structure may be developed in any
vitreous rock.

Many of the lavas here mentioned might be described as
Rhyolites, or lavas which contain from over 60 to 80 per cent. of
silica, and which consist chiefly of orthoclastic felspar (sanidine),
more or less quartz, occasionally tridymite, and usually some
vitreous matter, or they may be wholly vitreous. Such rocks
nearly always show fluxion structure. The term rhyolite includes

---

[*] Quart. Journ. Geol. Soc., Vol. XXXIII., p. 449. "On certain ancient
Devitrified Pitchstones and Perlites from the Lower Silurian District of Shropshire."

crystalline volcanic rocks, such as quartz-trachytes, and, in a sub-group, the "hyaline rhyolites" or vitreous lavas of similar chemical composition. None of those now occurring in England and Wales retain their original glassy character. Since the devitrification of hyaline rhyolites approximates to, or is identical in character with the crystallization of those rhyolites which have assumed a crystalline structure at the time of their solidification from fusion, it is often extremely difficult, in the absence of perlitic structure, to distinguish between the members of the two series. In the following descriptions the name rhyolite, although a bad one in some respects, will be employed in doubtful cases.

## SHROPSHIRE, &c.

In localities where volcanic outbursts have occurred in recent or comparatively late geological periods, as in the Liparis, and the Yellowstone District, passages from crystalline to hyaline rhyolites may be seen.

The presence of rocks of this class in England and Wales was not recognised until Mr. S. Allport demonstrated their existence in his admirable paper, already cited. Within a short time similar rocks were recognised by other observers in the Silurian districts of Wales, Westmorland, and Cumberland. They have since been found in the Shetlands by Professor A. Geikie and Messrs. Peach and Horne,[*] in rocks of Old Red Sandstone age. Devitrified rhyolites also occur in Jersey, and Professor Ch. de la Vallée Poussin has succeeded in correlating similar rocks occurring in Belgium with those already met with in England and Wales.[†]

This eminent petrologist has arrived at the conclusion that the rocks formerly described as eurites occurring at Grand-Manil, near Gembloux, are certainly rhyolites, and he considers that they have once been vitreous, in part. They are associated with carbonaceous schists containing *Climacograptus scalaris*, while other beds in the immediate vicinity are replete with fossils characteristic of the Caradoc or Bala series. In Professor Poussin's work numerous comparisons are made of these ancient Belgian rhyolites with those of England and Wales occurring on the same stratigraphical horizon, the microscopic characters of the former having been studied side by side with some of the most typical sections figured and described in the foregoing pages.

Among the oldest rhyolites occurring in this country may be mentioned those from the neighbourhood of Wellington, in Shropshire, the principal exposures being at Lea Rock and Lawrence Hill, and also at Charlton Hill, and Caer Caradoc, near Church Stretton. They are associated with quartzites which were

---

[*] Trans. R. Soc. Edin., Vol. XXXII., Pt. II., p. 359.

[†] "Les Anciennes Rhyolites, dites Eurites, de Grand-Manil." Bulletins de l'Académie royale de Belgique 3ᵐᵃ série. t. x. No. 8, 1885.

mapped by the Geological Survey as Caradoc Sandstone. Mr. Allport, however, in common with many other observers, believes these beds to be of considerably earlier date. Fragments of rocks which Professor Bonney regards as devitrified rhyolites occur in certain breccias in Charnwood Forest.* The rhyolite first recognised as such by the same author, which occurs in the vicinity of Llyn Padarn, in North Wales, is also considered by him as a lava of Pre Cambrian age.† Similar rocks have also been described by Professor Bonney as occurring N.W. of Llanddeiniolen, in Caernarvonshire, and at Brithdir, near Bangor, in an appendix to a paper by Professor T. McK. Hughes, " On the Pre-Cambrian Rocks of Bangor."‡ Mr. T. Davies, in an appendix to a paper by Dr. Hicks, has also described an agglomerate occurring at Caer-owdy, on the road from St. Davids to Solva, which consists largely of rhyolitic fragments.§ Professor Bonney, in an appendix to a paper by Dr. Callaway, records the occurrence of a trachyte at Carreg-winllan, Pensarn, S.E. of Amwlch, in Anglesey.‖ As the trachytes are closely related to the felsitic lavas this is worth noting here.

Leaving, now, these rocks, of which the precise age is doubtful, and of which notices will be found in the papers cited, we may turn to the description of similar lavas that were unquestionably erupted during the Silurian period.

## LAKE DISTRICT.

At RED CRAG, 1½ mile N.E. of Stockdale, Westmorland, there is a delicately banded rock which, under the microscope, shows very distinct perlitic fissures over the entire section (Pl. II., Fig 2). A finer example of this structure is seldom to be seen. The devitrification is spherulitic and crypto-crystalline, the former condition appearing to predominate, but the spherules are exceedingly minute and only to be detected on attentive examination. Some isotropic matter (? microfelsitic base) occurs in the section. The rock is a devitrified perlitic obsidian. It is of Lower Silurian age. The specimen here described was collected by Mr. Aveline.

At the northern end of the LONG SLEDDALE Valley, in Westmorland, and at several other places along the outcrop of the Coniston Limestone, a compact and often pinkish felsite occurs, which frequently shows a very delicate banding.

A specimen collected by the author at TILL'S HOLE, at the northern end of the Long Sleddale Valley, shows, under the microscope, a well-marked perlitic structure, and is completely devitrified, chiefly by the development of minute spherules with

---

* Quart. Journ. Geol. Soc., Vol. XXXIV., p. 207.
† Ib. Vol. XXXV., p. 309.
‡ Ib. Vol. XXXIV., p. 137.
§ Ib. Vol. XXXIV., p. 153.
‖ Ib., Vol. XXXVII., p. 236.

a radiating crystalline structure. That this was once a perlitic obsidian, or perlite, as some might prefer to call it, there can be no doubt. The section is traversed by delicate bands (fluxion structure) which are crossed by the perlitic cracks (Pl. II., Fig. 1). Spherulitic devitrification followed the development of the perlitic structure.

The rock from which this section was cut is of a pale pinkish-buff, or light yellowish-brown tint, traversed by delicate brown bands, like fine lines drawn in sepia and intersected by little veins of quartz.*

In the English Lake District other felsitic lavas of Lower Silurian age have also been collected by Mr. W. Talbot Aveline and Mr. Thos. Hart. Those collected by Mr. Aveline are chiefly from the district to the north-west of Conistou Lake; while Mr. Hart's specimens were derived from the neighbourhood of GRIZEDALE TARN. One of the latter, showing a well-marked columnar jointing, came from a spot 200 yards east of the tarn. The rock is of a dark ashen greenish-grey tint, the joint surfaces being chocolate-brown. The specimen is a somewhat irregular six-sided prism, terminated at one extremity by a flat joint plane which cuts the axis of the prism at an angle of about 15°.

In thin section, under the microscope, this rock shows well-defined perlitic structure. It was, therefore, evidently vitreous. The section is pervaded by a deep green tint, due to the presence of some mineral of secondary origin, probably chlorite. The occurrence of this rock, as a lava, is not indicated on the Survey Map; but the exposure, according to Mr. Hart, is not a large one.†

About three-quarters of a mile N.N.E. of SEATHWAITE CHURCH, IN THE DUDDON VALLEY, a rhyolitic rock has been mapped by Mr. Aveline, in which, under the microscope, a mottled streaking occurs. These markings are not very strong, but they clearly show the character of the rock. In the same section some cracks, much more strongly defined, traverse the direction of flow in straight and reticulating lines. The rock is of a dark brownish-green colour, with minute greyish flecks. It has a slightly greasy lustre and a somewhat conchoidal fracture.

In the COPPER-MINE VALLEY, N. OF CONISTON, a sage-green faintly-mottled felsitic rock occurs. It has a conchoidal fracture, and shows under the microscope a few indications of perlitic structure. The groundmass of this rock transmits very little light between crossed nicols. Numerous microliths and a few porphyritic crystals of felspar occur in the section. The latter are very clear and glassy. This, in conjunction with the uniform texture of the groundmass, the faint evidence of perlitic structure, and the conchoidal fracture of the specimen, indicate that it was once a vitreous rock.

---

* Quart. Journ. Geol. Soc., Vol. XL., Pl. XVIII.
† The columnar structure of the rock was noted by the late Mr. Ward on his MS. Maps.

A pale-drab felsite, very compact, and with an irregular or splintery fracture, occurs on THE KNOTT, BROUGHTON MOOR. It has a felsitic groundmass of fine texture, in which coarser isolated grains and aggregates of grains occur, either forming pseudomorphs after minute crystals, or wavy strings which follow a general direction. The banded structure is best seen by ordinary transmitted light, and it is very suggestive of fluxion structure. Some of the felsitic lavas constituting part of CONWAY MOUNTAIN, in North Wales, are very similar in microscopic character.

On the WESTERN SIDE OF GREAT STICKLE, the bottom bed of the series appears, from both field evidence and microscopic characters, to be a felsitic lava.

A felsite, which may or may not be a rhyolite, occurs at COLT CRAG, on the east side of Coniston Old Man.

The foregoing rocks will be described in the Memoirs of the Geological Survey, Explanation of Quarter-Sheet 98 N.W., in which certain rhyolite-tuffs are likewise noticed. These tuffs occur at the following localities :—

1. AT APPLETREEWORTH; in beds which, according to Mr. Aveline, lie conformably beneath the Coniston Limestone. This is an exceedingly good example of a rhyolite-tuff or obsidian-tuff.

2. S.W. OF CONISTON; in a bed lying conformably to and directly beneath the Coniston Limestone, and resting unconformably on the older volcanic series, according to Mr. Aveline.

3. At the WEST END OF THE YEWDALE BRECCIA; where it is lost below the Coniston Limestone, as stated by Mr. Aveline. The fragments in this tuff appear to be only partially rhyolitic.

A microscopic section made from a specimen collected by Mr. Aveline on the south side of the Beck, 300 yards N.W. OF SPA WELL, SHAP WELLS, shows very delicate and exceedingly convoluted banding, while in some parts crystallisation along certain lines has given rise to structures closely resembling the axiolites described by Zirkel. The section is partly cryptocrystalline, and contains quartz both in porphyritic crystals with irregular contours, in perfectly rounded crystals, and in crystals in which the angles are only slightly rounded. When magnified about fifty diameters parts of the section exhibit a damascened appearance like the mottle on a gun-barrel. There seems little doubt that the rock from which this specimen was derived is a felsitic lava.

Among the specimens collected by the late Rev. J. Clifton Ward is a felsite which passes "through the trap of KNOTT END." It consists of greyish-green and deep red or brownish-red bands and spots. Under the microscope one set of bands is seen to be partly devitrified by crystalline grains and partly by microliths while the others are much more transparent in thin section, have a deep yellow or orange colour and a more glassy appearance, and transmit far less light between crossed nicols. They are, however, devitrified to a considerable extent by microliths. These latter bands show distinct evidence of spherulitic structure, and may be regarded as more or less continuous belts of spherules. The rock is probably a devitrified spherulitic obsidian or pitch-stone.

## NORTH WALES.

The most numerous, and among them some of the best, examples of felsitic and devitrified lavas are met with in different parts of Wales.

Among the most striking peculiarities which these rocks present may be mentioned the occurrence in some of them of large spherical or spheroidal nodules. Bodies of this kind have already been alluded to on p. 8.

They are well developed in the felsite of ESGAIR-FELEN, a spur of the GLYDERS, overlooking the PASS OF LLANBERIS.

A specimen of a similar rock was collected by the author some years ago at BODLONDEB POINT, CONWAY. This is probably a continuation of the nodular felsite mentioned by Prof. Bonney as occurring in the DIGANWY HILLS on the opposite side of the Conway Estuary, and described by him in his paper, "On some Nodular Felsites in the Bala Group of North Wales."[*] In the same paper other nodular felsites have been noted between CONWAY FALLS and PANDY MILL, in the LEDR VALLEY, about half a mile above the new viaduct, where nodules from one to one and a half inches, and sometimes nearly three inches in diameter occur, and in the upper part, and at the western end, of Conway Mountain. In the last-named locality Professor Bonney states that he believes the rock still retains a glassy residuum, a point of interest in rocks of such great antiquity.

In addition to the spherulitic structures occurring in the lavas of ESGAIR-FELEN a distinctly perlitic one has been described by the author, thus leaving no doubt that the rock was once vitreous. In one section from this locality the perlitic structure is very curiously restricted to certain definite areas, Fig. 1, Pl. IV., the boundaries of which follow the contours of mosaic-like aggregates of quartz, which may be regarded as filled-up vesicles. This phenomenon is probably due to an originally different state of tension in the glass immediately surrounding the vesicles.

Another felsitic lava, which crops out about 100 feet from the base of the Glyder Fawr,[†] shows a small spherulitic structure under

---

* Quart. Journ. Geol. Soc., Vol. XXXVIII., p. 289.
† Ib., Vol. XXXV., p. 508.

the microscope, the spherules, or rather radiating crystallisations which appear to constitute the entire rock, having usually very jagged boundaries (*see* Fig. 4, Pl. II.).

In connexion with the larger spheroidal or nodular bodies it is worth noting that a thin section through a concretion or spheroid, about an inch in diameter, from one of the felsites of Skomer Island, off the west coast of Pembrokeshire, is traversed by slightly-flexed, approximately-parallel bands, some of them extremely narrow and charged with nearly opaque ferruginous matter, while other bands show a fine crypto-crystalline structure. The matter lying between these bands is coarsely crystalline, this texture contrasting strongly with that of the more delicate belts. These spheroids, evidently of secondary origin, may be compared with those occurring in the Magnesian Limestone of Durham, which, in a similar manner, are traversed by the planes of lamination in the limestone, while this Skomer Island concretion is traversed by the bands common in viscous lavas.

Close to the GLYDERS and on the right-hand side of the road ascending from PONT-Y-GROMLECH to GORPHWYSFA a dark-grey felsite occurs which has a fissile, platy structure, and a splintery fracture. When examined microscopically it shows great numbers of irregularly-shaped shreds and strings, (Pl. II. Fig. 6).

In polarised light the rock is seen to be felsitic throughout, but the crystalline texture of the shreds differs somewhat from that of the groundmass of the section. A striking similarity is seen between the microscopic structure of this rock and that of a Hungarian Obsidian, from Tolcsva. A rhyolite from Gardiner's River, Montana, U.S., also exhibits a similar structure.[*] It may be worthy of note that the forms of these shreds, as seen in section, bear a certain resemblance to the forms of shreds of glass met with in some volcanic ejectamenta. There seems, however, but little doubt that the Pont-y-Gromlech rock is a devitrified obsidian or rhyolite, similar to the Tolcsva or to the Gardiner's River rocks.[†] The structure is not by any means common, and appears to be due to the extremely convoluted bands being cut through in a given direction, so that only portions of the bands are included between the planes of section. What appears to be a similar structure, but on a much larger scale, occurring in a felsite on the Minnesota shore, is figured by Mr. R. D. Irving in his treatise " On the Copper-bearing Rocks of Lake Superior," where he attributes it to " flowing in a viscid condition."[‡]

---

[*] Quart. Journ. Geol. Soc. Vol. XXXVII., p. 406.
[†] Professor Ch. de la Vallée Poussin figures and describes a precisely similar rock, which occurs in beds approximately of Bala age at Grand-Manil, near Gembloux, in Brabant. " Les Anciennes Rhyolites, dites Eurites, de Grand-Manil." Bull. Acad. royale de Belg.. 3ᵐᵉ serie, t. x., No. 8, 1885.
[‡] Third Annual Report, U.S. Geol. Survey, p. 129.

A specimen of felsitic lava from CONWAY MOUNTAIN[*] shows a well-marked banded structure. Under the microscope it presents a very uniform felsitic appearance and shows some pseudomorphs of exceedingly fine-grained felsitic matter, after felspar crystals. These crystals or pseudomorphs lie with their longest axes parallel with the banding of the rock (Fig. 5., Pl. III.).

At CLOGWYN DÚR ARDDU, a ridge about a mile to the N.W. of the summit of SNOWDON, a well-banded devitrified obsidian occurs (Pl. II., Fig. 5), showing faint, though unmistakeable traces of perlitic structure, which on account of its faintness renders this section especially worthy of study.[†] It indicates, indeed, that perlitic structure *may* become completely effaced by devitrification, and that we may, therefore, often be justified in the assumption that many felsites which now show no trace of perlitic structure may once have been vitreous. Although we may look for this structure as a trustworthy test of the former vitreous character of a rock, yet we must not lose sight of the fact that there are vast numbers of specimens of recent obsidians and other glassy rocks which show no trace of perlitic structure, and it would therefore be not merely injudicious, but erroneous, to exclude all felsites, in which this structure is absent, from the category of vitreous rocks. On this account some examples of what are provisionally termed "doubtful felsitic lavas" are appended to this monograph. In connexion with this doubtful series it is interesting to note the close resemblance between some of the claystone porphyries (*Thonsteinporphyr*) and certain devitrified rhyolitic rocks. Apart from the devitrification which has taken place in the Clogwyn dúr Arddu lava, there is no structural distinction to be found between it and ordinary banded obsidians. The structure of rocks of this class may, however, be somewhat difficult to interpret at times. In a section of a banded felsitic lava, similar to that from Clogwyn dúr Arddu, but derived from Vicart Point, on the north coast of Jersey, the banding has been greatly disturbed by crushing and displacement, no less than a dozen minute faults occurring in one band within a space of less than an inch, while fragments of some bands abut against others at angles of about 45 degrees. The resemblance of the lavas of Clogwyn dúr Arddu to the obsidian of Ascension was long ago noted by Sir Andrew Ramsay.[‡]

On the south side of the Capel·Curig Road, about a quarter of a mile from BEDDGELERT, Mr. G. G. Butler collected specimens of a devitrified spherulitic obsidian or pitchstone.[§] The rock is dark greenish-grey, spotted with pale greenish-grey spherules, some isolated and over ¼ inch in diameter (Pl. II., Fig 3.), while others have coalesced, forming bands ⅛ to ¼ inch in breadth.

---

[*] First quarry past the turnpike on the road to Penmaen Mawr.
[†] This structure is not indicated in Fig. 5, Pl. II. It occurs in another part of the section, and could only be properly rendered by a more delicate style of engraving.
[‡] "Descriptive Catalogue of Rock Specimens," 3rd Edition, 1862, Mus. Pract. Geol.
[§] Quart. Journ. Geol. Soc., Vol. XXXVII., p. 404.

Under the microscope the spherules are seen to have narrow
but clear and colourless, borders, very sharply defined, while
similar sharply-defined lines traverse the matrix in which the
spherules lie. Some of these lines are straight, while others
describe curves which indicate a rude perlitic or spheroidal
structure. Perlitic markings on a smaller scale are also visible
in some parts of the section. Figures illustrative of these structures
are appended to the paper already cited.

On the eastern side of the EIFL RANGE, in Caernarvonshire, a
rhyolite occurs which has been described by Professor Bonney.[*]

## SKOMER ISLAND.

The felsitic lavas of Skomer Island off the West Coast of
Pembrokeshire and south of St. Bride's Bay, must be regarded as
of Lower Silurian age, as indicated by Sir Roderick Murchison.[†]
Specimens from this locality, collected by Sir Andrew Ramsay,
are deposited in the Rock Collection in the Museum of Practical
Geology, and it is from sections taken from some of them that the
following notes have been made.

The felsitic lavas of this island are probably the "quartzose
cornean, mostly striped," of Sir Henry Dela Beche,[‡] who also
mentions the occurrence of a "fine-grained compact greenstone—
sometimes approaching to cornean." A section made from one of
these "greenstones" in the Jermyn Street collection shows that it
is a basalt.

These rocks are considered to belong to the Llandeilo or to the
Bala series. Their microscopic characters were first described in
1881.[§] They show, as a rule, a well-marked banding; while
occasionally they contain spherules an inch or more in diameter.
Under the microscope both spherulitic and perlitic structures are
beautifully defined. The microscopic spherules generally have a
radiating crystalline structure, but in the larger ones this is either
imperfectly developed, so that they show no dark cross between
crossed nicols, or else it is absent. These larger bodies sometimes
interrupt the bands composed of the smaller spherules. In the
section from which Fig. 1. Pl. III., was drawn, by ordinary transmitted light, the finely-banded, perlitic and spherulitic structure of
the rock is admirably seen. The spherules appear colourless, but the
section is pervaded by clear brownish-green serpentine, seemingly
quite structureless, which lies between the perlitic cracks and commonly occupies the central area of each perlitic section, so that we
can only conclude that in most cases the nuclear portion of each

* Quart. Journ. Geol. Soc., Vol. XXXV., p. 305.
† "Siluria" 4th Edition, p. 144.
‡ Trans. Geol. Soc., 2nd Series, Vol. II., p. 8.
§ Quart. Journ. Geol. Soc., Vol. XXXVII., p. 409.

perlitic body was more readily soluble than the rest of the rock, and that after its removal serpentinous matter refilled the cavity and that similar infiltrations in percolating along the minute planes of perlitic and laminar, or banded, structure formed irregular deposits of serpentine, which substance represents, indeed, a large proportion of the rock in its present state. There is not the slightest indication that this material has resulted from the alteration of any magnesia-bearing mineral, so that we can only assume that it has been brought in solution from some other source. It may possibly be derived from basalt, which also occurs in the island. The curious way in which this green matter follows the minute structure of the rock, especially the perlitic, picking it out in a manner which renders the structure unusually striking, is a sufficient proof, if any be needed, of the secondary origin of this substance. The rock is throughout highly spherulitic, in fact spherules occur wherever serpentine does not occur. These are generally very small, having a radiating crystalline structure and showing dark crosses between crossed nicols. They appear massed together in somewhat irregular belts of variable width, and it is interesting to note that there is a very delicate banded structure passing through them; just such as one sees in modern vitreous lavas, which is perfectly distinct, in spite of the immense antiquity of the rock and the great changes which it has undergone.

There are, in addition to the little spherules which constitute so great a portion of the rock, some much larger ones. These often interrupt the bands, composed of the little spherules, quite abruptly. They show a tendency towards a radiate crystalline structure, but the crystallisation in these larger spherules is very irregular, and they show no interference-cross between crossed nicols. This is one of the most beautiful examples of a devitrified lava to be met with in rocks of Lower Silurian age. Another specimen from the same locality presents somewhat different microscopic characters, no perlitic structure being visible; but, by ordinary transmitted light, the section is seen to be traversed by irregular bands, which look like segregations of brownish or greenish-brown dust (Fig. 3, Pl. III.), with numerous circular or oval spots, which, under a power of about 250 diameters, appear to be segregations of very fine chloritic scales. It is evident, when the section is viewed between crossed nicols, that these spots mark the position of spherules, which in many cases have been obliterated; and we may regard these dusty-looking belts as spherulitic bands, while in some instances we may see unaltered bands composed of spherules, the chloritic matter having segregated only along the margins of these bands. The broader intermediate belts consist of crypto-crystalline matter with a few included spherules and irregularly-shaped pale green or greenish-brown plates of a chloritic mineral.

An additional example of what appears to be an altered spherulitic structure is afforded by the section shown in Fig. 2, Pl. III,

Here, by ordinary transmitted light, we find the rock traversed by approximately clear and colourless bands, irregularly moniliform, often with adherent circular patches of similarly clear, colourless matter ; while like circular areas, doubtless representing spherules, lie scattered between these clear bands. The intercalated bands consist of greenish-brown chloritic matter in intimate admixture with fine doubly-refracting particles. When the section is rotated between crossed nicols it is evident that if the clear bands and circular areas be sections through spherulitic belts and isolated spherules, they have, at all events, parted with their radiating crystalline structure ; and in the case of certain narrow moniliform belts, a strong jointed structure may be traced across them approximately at right-angles to the axis of the belt, the spaces between each pair of joints being occupied by perfectly distinct crystalline bundles, (Figs. 3 and 4, Pl. IV.). This structure has evidently been superinduced since the formation of the spherules, and is precisely analogous to that met with in the devitrification of thin, parallel-sided plates of artificially formed glass*. That we have in this, and the preceding rock, examples of devitrified spherulitic obsidians there can be little doubt.

## PEMBROKESHIRE.

NEAR FISHGUARD and at STRUMBLE HEAD, on the north coast of Pembrokeshire, certain felsitic rocks occur which have been mapped by the Geological Survey as "altered Llandeilo flags." Specimens so labelled, which have long been in the Rock Collection in the Museum of Practical Geology, seemed to bear so close a resemblance to devitrified lavas that sections of them were cut, and subsequent microscopic examination indicates that the rock from LLANWNDA, near Fishguard, is of a rhyolitic type.

It is crypto-crystalline, has a banded structure, and contains small lath-shaped felspar crystals, ranged with their longest axes approximately parallel to the banding (Fig. 6, Pl. III.).

In the specimen from STRUMBLE HEAD, collected between the Camp and the North Coast, the banding is absent, at all events in the section examined, yet although there is no proof that it is a rhyolitic rock, its microscopic character is not incompatible with such a supposition, while its occurrence on approximately the same horizon as the Llanwnda felsite lends support to the view that it may be an extension of it. Under the microscope it is seen to consist of a mesh-work or felting of delicate crystalline spiculæ, with apparently a little isotropic matter. It is a bluish-grey flint-like rock or hornstone. A specimen, also in the Jermyn Street collection, labelled, "Altered Llandeilo Flags from Carneddu, near Builth," presents a somewhat similar appearance under the micro-

---

* Figured in a paper "On the Microscopic Characters of some Specimens of Devitrified Glass, with notes on certain analogous Structures in Rocks." Hermann and Rutley. Proc. Royal Soc., Vol. XXXIX., p. 87.

scope, but, in this case, the section, when rotated between crossed nicols, breaks up into a series of irregularly rounded and angular patches, much more distinct than any in the Strumble Head section. An examination of more specimens from this locality seems needful before any decided opinion can be given concerning the original nature of this rock. It is, indeed, only by carefully studying the microscopic characters of a series of more or less altered slates and flags, and by comparing them with felsites of known origin and of different crystalline textures, that the difficulty of distinguishing the rocks of the one class from those of the other can be fully realised. It would, in fact, seem that we have passages from slates and flags into felsites, so that in the absence of any structural peculiarities it is scarcely possible to distinguish the one class of rocks from the other.

When we consider the nature of the minute particles of which slates are composed, a point most ably demonstrated by Dr. Sorby,[*] we need feel but little surprise at the occasional doubt which the microscopist experiences when dealing with rocks of this kind, while a comparison of hand specimens of felsitic lavas of the hornstone type with altered slates and porcellanites of similar aspect shows that, without the microscope, we may meet with like embarrassment.

Mr. T. Davies, in describing specimens from the TREFFGARN ROCKS, on the Fishguard Road, evinces great caution, stating that " the macroscopical and microscopical characters of these rocks are so remarkably like those of the hälleflintas from Sweden, that they are not to be differentiated by means of the microscope." He is inclined to regard them as of sedimentary origin.[†]

Closely allied to these rocks are certain spherulitic quartz-porphyries which, according to Professor A. Geikie, are probably of Lower Silurian age. They occur in the ST. DAVIDS' district, near NUN'S CHAPEL and CHURCH SCHOOL QUARRIES. They are intrusive in slaty beds associated with granite, and their resemblance to spherulitic rhyolites is very close, except in the coarser crystallisation of the materials of the groundmass. These rocks have been described by Professor A. Geikie in his paper, " On the supposed Pre-Cambrian Rocks of St. Davids,"[‡] to which microscopic drawings by Mr. F. W. Rudler (Figs. 7 and 8, Pl. X.) are appended. Mr. T. Davies has also described these and some other, felsitic rocks.[§] It seems quite possible that these spherulitic quartz-porphyries may represent the "feeders" from which some of the Silurian rhyolites emanated. They certainly appear to have a character intermediate between that of an ordinary quartz-porphyry and a quartz-rhyolite.

The quartz-rhyolite of LLYN PADARN has already been alluded to on page 12. It shows very distinct fluxion structure and

---

\* Quart. Journ. Geol. Soc., Vol. XXXVI., Presidential Address, 1880.
† Quart. Journ. Geol. Soc., Vol. XXXV., p. 292.
‡ Quart. Journ. Geol. Soc., Vol. XXXIX., p. 315.
§ Ib., Vol. XXXIV., p. 164, and Vol. XXXV., p. 293.

contains numerous crystalline grains of quartz, which sometimes include small fluid lacunæ. The rhyolitic character of this rock was first recognised by Professor Bonney.

## POST-SILURIAN FELSITIC LAVAS.

No felsitic lavas of Post-Silurian age have yet been recorded in England or Wales, with the exception of one occurring at BRENT TOR, in Devonshire,* described by the author some years ago as a devitrified rhyolite or rhyolitic breccia showing delicate sinuous bands which appear to indicate fluxion structure, but the rock, like most of those in the vicinity, is much altered and infiltered with silica, and the evidence of its rhyolitic character is obscure. It contains numerous opaque scoriaceous lapilli, and is either of late Devonian or early Carboniferous age, probably the former. Several other banded felsitic rocks, also from Devon, which closely simulate felsitic lavas, have been cut and examined microscopically, but they have proved to be merely altered sedimentary deposits. A rock from the Mendips, described long ago by the author as a "devitrified pitchstone porphyry,"† is simply a porphyrite with some devitrified matter in the groundmass. In the SHETLANDS, true rhyolites of Old Red Sandstone age have been met with by Professor A. Geikie and Messrs. Peach and Horne.‡

Felsitic lavas, both of the Old Red Sandstone period and also of Tertiary age, have been noted by Professor Judd§ as occurring in the district of LORNE and in the ISLE of MULL. He states that in their weathered condition they sometimes show a spherulitic or perlitic structure. Furthermore, truly vitreous lavas of Tertiary age occur in the island of EIGG, in ARDNAMURCHAN, &c., while the well-known pitchstones of ARRAN are probably referable to the same period.

Porphyrites, sometimes with a vitreous base, are met with in the extreme North of England. As these cannot be classed with the lavas which form the subject of this monograph, and since the rhyolites of Old Red Sandstone and later age lie outside the boundaries of England and Wales, they can here claim merely a passing notice.

## FELSITES OF DOUBTFUL ORIGIN.

A felsite from the WEST OF LLYN ARENIG, six miles west of Bala, appears from its microscopic characters to be a devitrified, banded obsidian (Pl. III., Fig. 4). It has a very uniform crypto-crystalline structure. The bands are of finer texture than the

---

* Memoirs of the Geol. Survey of England and Wales, "The Eruptive Rocks of Brent Tor," p. 32. Pl. VII., Fig. 1.

† Mem. Geol. Survey, "Geology of E. Somerset and the Bristol Coal-fields"; Appendix I., p. 208; Pl. VII.

‡ Trans. Royal Soc., Edin., Vol. XXXII., part 2.

§ Quart. Journ. Geol. Soc., Vol. XXX., p. 220.

rest of the rock. It contains a few small felspar crystals, some of which are triclinic. No perlitic structure is visible, so that there is no *proof* that it was ever a vitreous rock.

A felsite from MOEL-Y-MENYN, two miles S.S.E. of Arenig, shows no microscopic structures which permit the formation of a definite conclusion as to its origin. The same might be said of a felsite from the east side of Y. Graig, about eight miles S.W. of Bala, except that in this case fragments of other rocks and broken felspar crystals appear to be present, so that it is probably a tuff. A section made from a specimen from the top beds on the EAST SLOPE OF ARENIG, in Merionethshire, affords no satisfactory microscopic evidence as to its origin. It is essentially a felsite, and may have been either a very fine-grained volcanic ash or a lava. By ordinary light it shows peculiar markings rather like those in the Pont-y-Gromlech rock (p. 16). Irregular spots of green matter are present, which, from their form, may be pumice fragments, filled with chlorite*, but there is no good proof of this. No doubt an examination of more specimens from the same beds might give a clue one way or the other.

The felsite of MYNYDD NODOL, 5 miles W.N.W. of Bala, is a rock of much the same microscopic character as the preceding, but has rather more the aspect of a lava. There seem to be indications of fluxion structure, but they are very obscure.

A felsite which, under the microscope, shows a considerable number of porphyritic quartz crystals around which sweep films of a chloritic or sericitic mineral, occurs at Y-GARN, near Llanrhaidr-yn-Mochnant. It may be a quartz-rhyolite.

Between TAN-Y-GRISIAN and CWM ORTHIN, near FFESTINIOG, and at CAREGLWYD, LLANFECHELL, ANGLESEY, felsites of doubtful origin, so far as microscopic evidence goes, also occur.

Many other felsites might be mentioned concerning which it is difficult or impossible to give any decided opinion from their microscopic characters. Some may be lavas, and some altered tuffs; while others may be porcellanites, or less metamorphosed slates and shales, and, at times, even fine-grained felspathic grits which have undergone more or less change may, from their mineral constitution, be included in the somewhat elastic group of felstones.

A very good example of the obliteration of structural peculiarities in rocks of this class is seen in a microscopic section of a little felsitic vein, a friction-breccia about five inches broad, passing through fine-grained Silurian grits in Smooth Beck, N.E. of ESTHWAITE WATER, in Westmorland, in which by ordinary transmitted light the forms of angular and rounded fragments are shown only by pale brown, nearly washed-out-looking patches, which, between crossed nicols, are scarcely to be distinguished from their felstic cement. Similar effacement of fragmental

---

* *See* description of pumice fragment by Sorby. Presidential Address, Quart. Journ. Geol. Soc., Vol. XXXVI., p. 80, 1880.

structure may also be noted in highly altered volcanic agglo-merates.

So deceptive, sometimes, is the aspect of altered clastic rocks, that it is scarcely possible to distinguish by the eye alone between them and felsitic lavas, especially when the specimens possess well-marked lamination and the surfaces have been somewhat weathered.

The foregoing pages probably give only a few of the English and Welsh localities where felsitic lavas occur, although it is hoped that most of the spots where they are at present known have here been recorded. No doubt, by following along the strike of felsitic beds now known to be lavas, the list of localities might be added to considerably; while further examination of felsites of doubtful origin in the field may lead to the recognition of many old lava-flows which have hitherto passed unnoticed. In doubtful cases the evidence gathered in the field is quite as valuable as, or more so than, that derived from microscopic investigation.

| | Quartz-Trachyte, Rhyolite, Liparite | Sanidine-Trachyte, Sanidine-Rhyolite | Trachyte | |
|---|---|---|---|---|
| | can pass. Contains little or no water. Silica = 60 to 80 %. | | | V<br>O<br>L<br>C<br>A<br>N<br>I<br>C |
| Crystalline. Commonly with more or less glassy matter. | *Quartz-Trachyte.* Rhyolite of Richthofen. Liparite of Roth. Silica = 75 to 77 %. | *Sanidine-Trachyte.* Sanidine-Rhyolite. Silica = 74 to 78 %. | *Trachyte.* Quartzless-Trachyte. Domite. Silica = 62 to 68 %. | |
| Crystalline. | | *Quartz-Porphyry.* Quartz-Felsite. Elvan. | | P<br>L<br>U<br>T<br>O<br>N<br>I<br>C |
| Crystalline. | | *Granite.* | | |

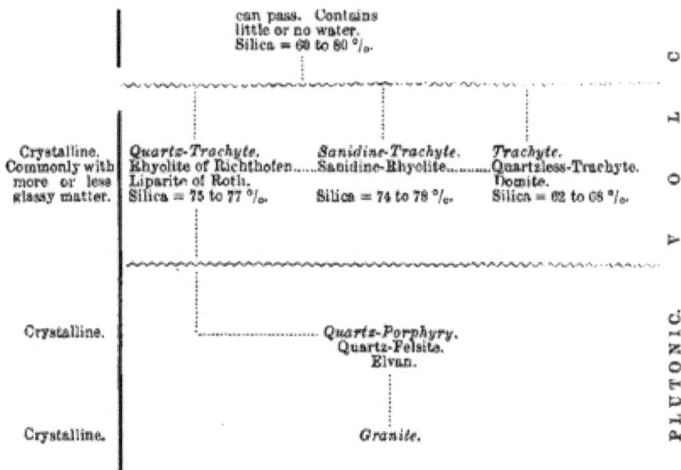

The following brief statement of the lithological characters of the rocks mentioned may help to render the preceding table more generally intelligible : —

*Felsite.*—Micro-crystalline, crypto-crystalline, micro-felsitic. Seldom any true glass.

*Obsidian.*—Glass containing crystallites and occasionally crystals. Banded, perlitic, and spherulitic structures common.

*Pitchstone.*—Glass with vast quantities of crystallites and often many porphyritic crystals. Banded, perlitic, and spherulitic structures common.

*Quartz-Trachyte, Rhyolite, or Liparite.*—Micro-crystalline or aphanitic ground-mass, occasionally micro-granitic. Porphyritic crystals of quartz and sanidine, also triclinic felspars, hornblende and biotite.

*Sanidine-Trachyte, Sanidine-Rhyolite.*—Micro-crystalline or aphanitic ground-mass, chiefly composed of felspars. Porphyritic felspars and biotite. Tridy-mite often present.

*Trachyte, Quartzless Trachyte, Domite.*—Crystalline groundmass generally, consisting chiefly of felspar microliths with a very small quantity of glass. Porphyritic sanidine, and often some triclinic felspars, hornblende, and biotite. High per-centage of silica usually regarded as due to tridymite.

*Quartz-Porphyry, Quartz-Felsite, Elvan.*—Micro-crystalline (felsitic) matrix. Porphyritic crystals of quartz and orthoclase. Mica seldom present, and then only in small quantity.

*Granite.*—Crystalline (completely). Essentially composed of crystals of quartz, felspar (chiefly orthoclase) and mica.

Obsidians and pitchstones occur both as lava flows and as dykes.

The trachytes occur chiefly as lava flows.

The quartz-porphyries form dykes.

Granite occurs in large intrusive masses and in veins.

# ALPHABETICAL LIST OF LOCALITIES.

# EXPLANATION OF THE PLATES.

## PLATE I.

Illustrative of structures occurring in vitreous and devitrified rocks.

The figures in this plate are diagrammatic, and represent thin sections, as seen under various magnifying powers, ranging from 10 to 50 linear.

Fig. 1. Spherule, with radiating crystalline structure and vitreous border.

" 2. Spherule, with similar structure to No. 1, but with a broad border formed of globulites.

" 3. Spherules, with radiating crystalline structure, but without borders—
  a. As seen by ordinary-transmitted light.
  b. As seen in polarised light between crossed nicols-prisms.

" 4. Spherulitic body, with radiating crystalline structure and irregularly indented surface. Not a spherule in form, but with the same radiating crystalline structure common in spherules.

" 5. Spherule, with granular structure.

" 6. Globulites segregating, so as to form what appears to be an incipient spherule.

" 7. Bands composed of microliths, deflected by a larger crystal (fluxion structure).

" 8. Bands, showing different degrees of fineness in their crystalline structure. Such bands may be micro or crypto-crystalline.

" 9. Bands composed of small spherules, as seen between crossed nicols.

" 10. Intricately convoluted bands (damascene structure).

" 11. Moniliform band, formed by the coalescence of spherules.

" 12. Parallel-sided band, cylindrical or tabular, formed by the coalescence of spherules.

" 13. Axiolites.

" 14. Perlitic structure.

## PLATE II.

" 1. Devitrified perlitic obsidian. Till's Hole. N. end of Long Sleddale Valley, Westmorland. Devitrified by spherules, which are only appreciable in polarised light. bb. Banding traversed by perlitic cracks. qq. Quartz-veins. × 32.

" 2. Devitrified perlitic obsidian. Red Crag, 1½ mile N.E. of Stockdale, Westmorland. Shows perlitic structure, and is devitrified partially by spherules. × 18.

" 3. Devitrified spherulitic pitchstone or obsidian. Beddgelert, N. Wales. Coarse spherulitic structure. Perlitic structure present, but not shown in drawing. × 18.

" 4. Spherulitic lava. Glyder Fawr, N. Wales. Made up of irregularly shaped spherulitic bodies. × 77.

" 5. Devitrified obsidian. Clogwyn dûr Arddu, Snowdon. Roughly parallel banding. Very faint perlitic structure present, but not shown in drawing. × 20.

" 6. Devitrified obsidian or rhyolite. Between Pont y-Gromlech and Gorphwysfa, Pass of Llanberis. Sections through convoluted bands. × 77.

# PLATE III.

Fig. 1, 2, and 3. Devitrified perlitic and spherulitic obsidians or pitchstones. Skomer Island, off the West Coast of Pembrokeshire. 1 and 2, × 18 : 3, × 55.

„ 4. Rhyolite or devitrified obsidian? W. of Llyn Arenig, 6 miles W. of Bala. × 18.

„ 5. Rhyolite or devitrified obsidian. Conway Mountain, N. Wales. × 18.

„ 6. Devitrified obsidian or rhyolite. Near Llanwnda, Fishguard, Pembrokeshire. × 18.

# PLATE IV.

„ 1. Devitrified perlitic obsidian. Esgair-felen, Y. Glyder Fawr, North Wales. × 10.

$FF$ = Felsitic matter like $PP$ between crossed nicols.
$PP$ = Areas in which perlitic structure occurs.
$qq$ = Aggregates of quartz.

„ 2. String of spherules in devitrified obsidian. Skomer Island. Ordinary illumination.

„ 3. The same in polarised light, showing transverse joints, separating distinct crystalline bundles.

PLATE I.

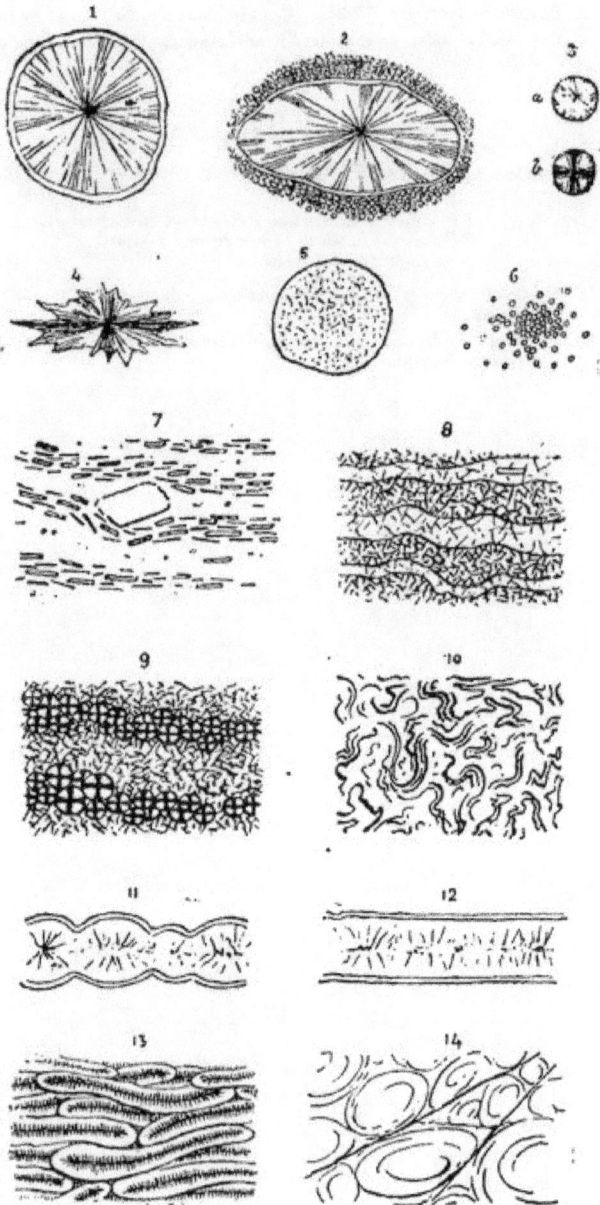

# PLATE II.

Fig. 1.

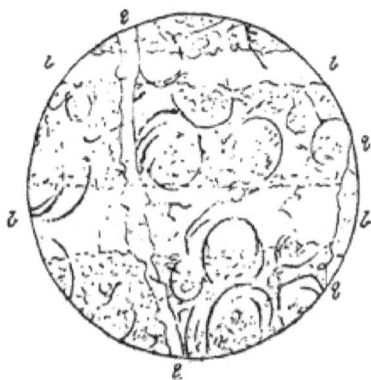

Fig. 2.

Fig. 3.

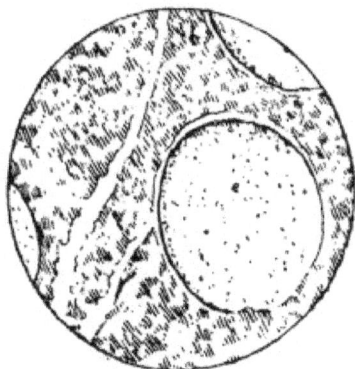

Fig. 4.

Fig. 5.

Fig. 6.

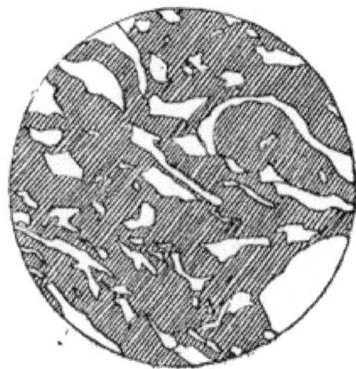

# PLATE III.

Fig. 1.

Fig. 2.

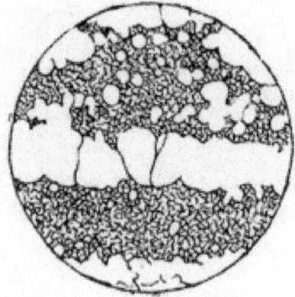

Fig. 3.

Fig. 4.

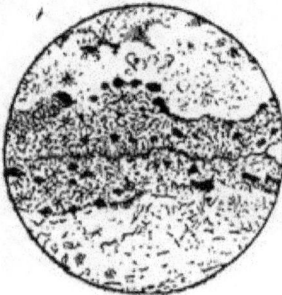
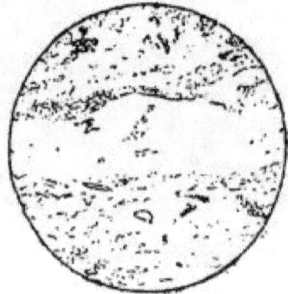

Fig. 5.

Fig. 6.

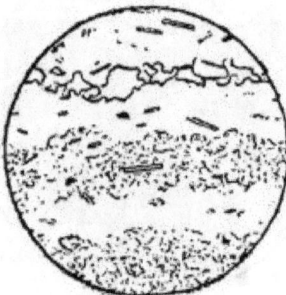

PLATE IV.

Fig. 1.

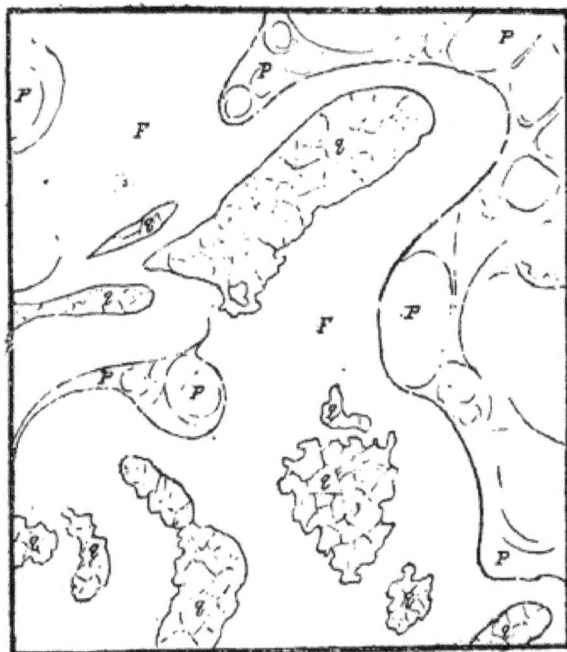

Fig. 2.

Fig. 3.

LONDON: Printed by EYRE and SPOTTISWOODE,
Printers to the Queen's most Excellent Majesty.
For Her Majesty's Stationery Office.
[8917.—375.—12/85.]

The CARBONIFEROUS LIMESTONE, YOREDALE ROCKS and MILLSTONE GRIT of N. DERBYSHIRE. By A. H. GREEN, DR. C. LE NEVE FOSTER, and J. R. DAKYNS. (2nd Ed. in preparation.)

The BURNLEY COAL FIELD. By E. HULL, J. R. DAKYNS, R. H. TIDDEMAN, J. C. WARD, W. GUNN, and C. E. DE RANCE. 12s.

The YORKSHIRE COALFIELD. By A. B. GREEN, J. R. DAKYNS, J. C. WARD, C. FOX-STRANGWAYS, W. H. DALTON, R. RUSSELL, and T. V. HOLMES. 42s.

The EAST SOMERSET and BRISTOL COALFIELDS. By H. B. WOODWARD. 18s.

The SOUTH STAFFORDSHIRE COAL-FIELD. By J. B. JUKES. (3rd Edit.) (*Out of print.*) 3s. 6d.

The WARWICKSHIRE COAL-FIELD. By H. H. HOWELL. 1s. 6d.

The LEICESTERSHIRE COAL-FIELD. By EDWARD HULL. 3s.

ERUPTIVE ROCKS of BRENT TOR. By F. RUTLEY. 15s. 6d.

FELSITIC LAVAS of ENGLAND and WALES. By F. RUTLEY.

HOLDERNESS. By C. REID. 4s.

BRITISH ORGANIC REMAINS. DECADES I. to XIII., with 10 Plates each. Price 4s. 6d. each 4to; 2s. 6d. each 8vo.

MONOGRAPH I. On the Genus PTERYGOTUS. By T. H. HUXLEY, and J. W. SALTER. 7s.

MONOGRAPH II. On the Structure of the BELEMNITIDÆ. By T. H. HUXLEY. 2s. 6d.

MONOGRAPH III. On the CROCODILIAN REMAINS found in the ELGIN SANDSTONES. By T. H. HUXLEY. 14s. 6d

MONOGRAPH IV. On the CHIMÆROID FISHES of the British Cretaceous Rocks. By E. T. NEWTON. 5s

The VERTEBRATA of the FOREST BED SERIES of NORFOLK and SUFFOLK. By E. T. NEWTON. 7s. 6d.

CATALOGUE of SPECIMENS in the Museum of Practical Geology, illustrative of British Pottery and Porcelain. By Sir H. DE LA BECHE and TRENHAM REEKS. 165 Woodcuts. 2nd Ed. by T. REEKS and F. W. RUDLER. 1s. 6d.; 2s. in boards.

A DESCRIPTIVE GUIDE to the MUSEUM of PRACTICAL GEOLOGY, with Notices of the Geological Survey the School of Mines, and the Mining Record Office. By ROBERT HUNT and F. W. RUDLER. 6d. (3rd Ed.)

A DESCRIPTIVE CATALOGUE of the ROCK SPECIMENS in the MUSEUM of PRACTICAL GEOLOGY. By A. C. RAMSAY, H. W. BRISTOW, H. BAUERMAN, and A. GEIKIE. 1s. (3rd Edit.) (*Out of print.*) 4th Ed. in progress.

CATALOGUE of the FOSSILS in the MUSEUM of PRACTICAL GEOLOGY:
CAMBRIAN and SILURIAN, 2s. 6d.; CRETACEOUS, 2s. 9d.; TERTIARY and POST-TERTIARY, 1s. 8d.

## SHEET MEMOIRS OF THE GEOLOGICAL SURVEY.

Those marked (O.P.) are Out of Print.

| 4 | · | · | FOLKESTONE and RYE. By F. DREW. 1s. |
|---|---|---|---|
| 7 | · | · | PARTS of MIDDLESEX, &c. By W. WHITAKER. 2s. (O.P.) |
| 10 | · | · | TERTIARY FLUVIO-MARINE FORMATION of the ISLE of WIGHT. By EDWARD FORBES. 5s. |
| 10 | · | · | The ISLE of WIGHT. By H. W. BRISTOW. 6s. (O.P.) |
| 12 | · | · | S. BERKSHIRE and N. HAMPSHIRE. By H. W. BRISTOW and W. WHITAKER. 3s. (O.P.) |
| 13 | · | · | PARTS of OXFORDSHIRE and BERKSHIRE. By E. HULL and W. WHITAKER. 3s. (O.P.) |
| 34 | · | · | PARTS of WILTS. and GLOUCESTERSHIRE. By A. C. RAMSAY, W. T. AVELINE, and E. HULL. 8d. |
| 44 | · | · | CHELTENHAM. By E. HULL. 2s. 6d. |
| 45 | · | · | BANBURY, WOODSTOCK, and BUCKINGHAM. By A. H. GREEN. 2s. |
| 45 SW. | | | WOODSTOCK. By E. HULL. 1s. |
| 47 | · | · | N.W. ESSEX & N.E. HERTS. By W. WHITAKER, W. H. PENNING, W. H. DALTON, & F. J. BENNETT. 3s. 6d. |
| 48 SW | | | COLCHESTER. By W. H. DALTON. 1s. 6d. |
| 48 SE | | | EASTERN END of ESSEX (WALTON NAZE and HARWICH). By W. WHITAKER. 9d. |
| 48 NW, NE. | | | IPSWICH, HADLEIGH, and FELIXSTOW. By W. WHITAKER, W. H. DALTON, and F. J. BENNETT. 2s. |
| 49 S and | | | ALDBOROUGH, FRAMLINGHAM, ORFORD, and WOODBRIDGE. By W. H. DALTON. Edited, with |
| 50 SE | · | } | additions by W. WHITAKER.) |
| 50 SW | · | · | STOWMARKET. By W. WHITAKER, F. J. BENNETT, and J. H. BLAKE. 1s. |
| 50 NW | · | · | DISS, EYE, &c. By F. J. BENNETT. 2s. |
| 51 SW · | | | CAMBRIDGE. By W. H. PENNING and A. J. JUKES-BROWN. 4s. 6d. |
| 53 SE · | | | PART of NORTHAMPTONSHIRE. By W. T. AVELINE and RICHARD TRENCH. 8d. |
| 53 NE · | | | PARTS of NORTHAMPTONSHIRE and WARWICKSHIRE. By W. T. AVELINE. 8d. (O.P.) |
| 63 SE · | | | PART of LEICESTERSHIRE. By W. TALBOT AVELINE, and H. H. HOWELL. 8d. (O.P.) |
| 64 · | | | RUTLAND, &c. By J. W. JUDD. 12s. 6d. |
| 65 NE, SE | | | NORWICH. By H. B. WOODWARD. 7s. |
| 66 SW · | | | ATTLEBOROUGH. By F. J. BENNETT. 1s. 6d. |
| 68 E · | | | CROMER. By C. REID. 6s. |
| 68 NW, SW. | | | PAKENHAM, WELLS, &c. By H. B. WOODWARD. 2s. |
| 70 | · | · | SW PART of LINCOLNSHIRE, with PARTS of LEICESTERSHIRE and NOTTS. By A. J. JUKES-Brown and W. H. DALTON. 4s. |
| 71 NE · | | | NOTTINGHAM. By W. T. AVELINE. (2nd Ed.) 1s. |
| 79 NW · | | | RHYL, ABERGELE, and COLWYN. By A. STRAHAN. (Notes by R. H. TIDDEMAN.) 1s. 6d. |
| 80 NW · | | | PRESCOT, LANCASHIRE. By E. HULL. (3rd Ed. With additions by A. STRAHAN.) - 3s. |
| 80 NE · | | | ALTRINCHAM, CHESHIRE. By E. HULL. 8d. (O.P.) |
| 80 SW · | | | CHESTER. By A. STRAHAN. 2s. |
| 81 NW, SW. | | | STOCKPORT, MACCLESFIELD, CONGLETON, & LEEK. By E. HULL and A. H. GREEN. 4s. |
| 82 SE · | | | PARTS of NOTTINGHAMSHIRE and DERBYSHIRE. By W. T. AVELINE. (2nd Ed.) 6d. |
| 82 NE · | | | PARTS of NOTTINGHAMSHIRE, YORKSHIRE, and DERBYSHIRE. By W. T. AVELINE. 8d. |
| 87 NW · | | | PARTS of NOTTS, YORKSHIRE, and DERBYSHIRE. (2nd Ed.) By W. T. AVELINE. 6d. |
| 87 SW · | | | BARNSLEY. By A. H. GREEN. 9d. |
| 88 SW · | | | OLDHAM. By E. HULL. 2s. |
| 88 SE · | | | PART of the YORKSHIRE COAL-FIELD. By A. H. GREEN, J. R. DAKYNS, and J. C. WARD. 1s. |
| 88 NE · | | | DEWSBURY, HUDDERSFIELD, and HALIFAX. By A. H. GREEN, J. R. DAKYNS, J. C. WARD, and R. RUSSELL. 6d. |
| 89 SE · | | | BOLTON, LANCASHIRE. By E. HULL. 2s. |
| 89 SW · | | | WIGAN. By EDWARD HULL. (2nd Ed.) 1s. (O.P.) |
| 90 SE · | | | The COUNTRY between LIVERPOOL and SOUTHPORT. By C. E. DE RANCE. 3d. (O.P.) |
| 90 NE · | | | SOUTHPORT, LYTHAM, and SOUTH SHORE. By C. E. DE RANCE. 6d. |
| 91 SW · | | | The COUNTRY between BLACKPOOL and FLEETWOOD. By C. E. DE RANCE. 6d. |
| 91 NW · | | | SOUTHERN PART of the FURNESS DISTRICT in N. LANCASHIRE. By W. T. AVELINE. 6d. |
| 92 SE · | | | BRADFORD and SKIPTON. By J. R. DAKYNS, C. FOX-STRANGWAYS, R. RUSSELL, and W. H. DALTON. 6d. |
| 93 NW · | | | NORTH and EAST of HARROGATE. By C. FOX-STRANGWAYS. 8d. |
| 93 NE · | | | The COUNTRY between YORK and MALTON. By C. FOX-STRANGWAYS. 1s. 6d. |
| 93 SW · | | | CARBONIFEROUS ROCKS N. and E. of LEEDS, and the PERMIAN and TRIASSIC ROCKS about TADCASTER. By W. T. AVELINE, A. H. GREEN, J. R. DAKYNS, J. C. WARD, and R. RUSSELL. 6d. (O.P.) |
| 94 NE · | | | BRIDLINGTON BAY. By J. R. DAKYNS and C. FOX-STRANGWAYS. 1s. |
| 95 SW, SE · | | | SCARBOROUGH and FLAMBOROUGH HEAD. By C. FOX-STRANGWAYS. 1s. |
| 95 NW · | | | WHITBY and SCARBOROUGH. By C. FOX-STRANGWAYS and G. BARROW. 1s. 6d. |
| 96 SE · | | | NEW MALTON, PICKERING, and HELMSLEY. By C. FOX-STRANGWAYS. 1s. |

96 NE - · ESKDALE, ROSEDALE, &c. By C. Fox-Strangways, C. Reid, and G. Barrow. 1s. 6d.
96 NW, SW NORTHALLERTON and THIRSK. By C. Fox-Strangways, A. G. Cameron, and G. Barrow.
98 SE - KIRKBY LONSDALE and KENDAL. By W. T. Aveline, T. McK. Hughes, and R. H. Tiddeman. 2s.
98 NE - KENDAL, WINDERMERE, SEDBERGH, & TEBAY. By W. T. Aveline & T. McK. Hughes. 3d. (O.P.)
101 SE - NORTHERN PART of the ENGLISH LAKE DISTRICT. By J. C. Ward. 9s.
104 SW, SE CLEVELAND. By G. Barrow.
108 SE · OTTERBURN and ELSDON. Hugh Miller. (Notes by C. T. Clough.)

## THE MINERAL DISTRICTS OF ENGLAND AND WALES ARE ILLUSTRATED BY THE FOLLOWING PUBLISHED MAPS OF THE GEOLOGICAL SURVEY.

### COAL-FIELDS OF ENGLAND AND WALES.

Scale, one inch to a mile.

Anglesey, 78 (SW).
Bristol and Somerset, 19, 35.
Coalbrook Dale, 61 (NE & SE).
Clee Hill, 55 (NE, NW).
Flintshire and Denbighshire, 74 (NE & SE), 79 (NE, SE).
Derby and Yorkshire, 71 (NW, NE, & SE), 82 (NW & SW),
81 (NE), 87 (NE, SE), 88 (SE).
Forest of Dean, 43 (SE & SW).
Forest of Wyre, 61 (SE), 55 (NE).
Lancashire, 80 (NW), 81 (NW), 89, 88 (SW, NW).
Leicestershire, 71 (SW), 63 (NW).
Northumberland & Durham, 103, 105, 106 (SE), 109 (SW, SE).
N. Staffordshire, 72 (NW), 72 (SW), 73 (NE), 80 (SE), 81 (SW).
S. Staffordshire, 54 (NW), 62 (SW).
Shrewsbury, 60 (NE), 61 (NW & SW).
South Wales, 36, 37, 38, 40, 41, 42 (SE, SW).
Warwickshire, 52 (NE SE), 63 (NW SW), 54 (NE), 53 (NW).
Yorkshire, 88 (NE, SE), 87 (SW), 92 (SE), 93 (SW).

### GEOLOGICAL MAPS.

Scale, six inches to a mile.

The Coal-fields and other mineral districts of the N. of England are published on a scale of six inches to a mile, at 4s. to 8s. each. MS. Coloured Copies of other six-inch maps, not intended for publication, are deposited for reference in the Geological Survey Office, 28, Jermyn Street, London.

#### Lancashire.

| Sheet. | Sheet. |
|---|---|
| 15. Ireleth. | 84. Ormskirk, St. Johns, &c. |
| 16. Ulverstone. | 85. Standish, &c. |
| 17. Cartmel. | 86. Adlington, Horwick, &c. |
| 22. Aldingham. | 87. Bolton-le-Moors. |
| 47. Clitheroe. | 88. Bury, Heywood. |
| 48. Colne, Twiston Moor. | 89. Rochdale, &c. |
| 49. Laneshaw Bridge. | 92. Bickerstaffe. |
| 55. Whalley. | 93. Wigan, Up Holland, &c. |
| 56. Haggate. | 94. WestHoughton, Hindley. |
| 57. Winewall. | 95. Radcliffe, Peel Swinton. |
| 51. Preston. | 96. Middleton, Prestwich. |
| 62. Balderstone, &c. | 97. Oldham, &c. |
| 63. Accrington. | 100. Knowsley, Rainford, &c. |
| 64. Burnley. | 101. Billinge, Ashton, &c. |
| 65. Stiperden Moor. | 102. Leigh, Lowton. |
| 69. Layland. | 103. Ashley, Eccles. |
| 70. Blackburn, &c. | 104. Manchester, Salford, &c. |
| 71. Haslingden. | 105. Ashton-under-Lyne. |
| 72. Cliviger, Bacup, &c. | 106. Liverpool, &c. |
| 73. Todmorden. | 107. Prescott, Huyton, &c. |
| 77. Chorley. | 108. St. Helen's, Burton Wood. |
| 78. Bolton-le-Moors. | 109. Winwick, &c. |
| 79. Entwistle. | 111. Cheadle, Stockport, &c. |
| 80. Tottington. | 112. Stockport, &c. |
| 81. Wardle. | 118. Part of Liverpool, &c. |

#### Durham.

| | |
|---|---|
| 1. Ryton. | 5. Greenside. |
| 2. Gateshead. | 6. Winlaton. |
| 3. Jarrow. | 7. Washington. |
| 4. S. Shields. | 8. Sunderland |

### Durham—*continued.*

| Sheet. | Sheet. |
|---|---|
| 9. | 25. Wolsingham. |
| 10. Edmondbyers. | 26. Brancepeth. |
| 11. Ebchester. | 30. Benny Seat. |
| 12. Tantoby. | 32. White Kirkley. |
| 13. Chester-le-Street. | 33. Hamsterley. |
| 16. Hunstanworth. | 34. Whitworth. |
| 17. Waskerley. | 38. Maize Beck. |
| 18. Muggleswink. | 41. Cockfield. |
| 19. Lanchester. | 42. Bishop Auckland. |
| 20. Hetton-le-Hole. | 46. Hawksley Hill House. |
| 22. Wear Head. | 52. Barnard Castle. |
| 23. Eastgate. | 53. Winston. |
| 24. Stanhope. | |

### Northumberland.

| | |
|---|---|
| 44. Rothbury. | 80. Cramlington. | 98. Walker. |
| 45. Longframlington. | 81. Earsdon. | 101. Whitfield. |
| | 82. NE. of Gilsland. | 102. Allendale |
| 46. Broomhill. | 83. Coadley Gate. | Town. |
| 47. Coquet Island. | 87. Heddon. | 103. Slaley. |
| 54. Longhorsley. | 88. Long Benton. | 105. Newlands. |
| 55. Ulgham. | 89. Tynemouth. | 106. Blackpool Br. |
| 56. Druridge Bay. | 91. Greenhead. | 107. Allendale. |
| 63. Netherwitton. | 92. Haltwhistle. | 108. Blanchland. |
| 64. Morpeth. | 93. Haydon Bridge. | 109. Shotleyfield. |
| 65. Newbiggin. | 94. Hexham. | 110. Wellhope. |
| 72. Redlington. | 95. Corbridge. | 111. Altonheads. |
| 73. Blyth. | 96. Horsley. | 112. |
| | 97. Newcastle. | |

### Cumberland.

| | |
|---|---|
| 55. Scarness. | 69. Buttermere. |
| 56. Skiddaw. | 70. Grange. |
| 63. Thackthwaite. | 71. Helvellyn. |
| 64. Keswick. | 74. Wastwater. |
| 65. Dockraye. | 75. Stonethwaite Fell. |

### Westmorland.

| | |
|---|---|
| 2. Tees Head. | 12. Patterdale. | 25. Grasmere. |
| 6. Dufton Fell. | 18. Near Grasmere. | 38. Kendal. |

### Yorkshire.

| | |
|---|---|
| | 116. Conistone | 260. Honley. |
| | Moor. | 261. Kirkburton. |
| 7. Redcar. | 135. Kirkby | 262. Dartoo. |
| 8. —— | Malham. | 263. Hemsworth. |
| 12. Bowes. | 184. Dale End. | 264. Cumpsall. |
| 13. Wycliffe. | 185. Kildwick. | 272. Holmfirth. |
| 20. Lythe. | 200. Keighley. | 273. Penistone. |
| 24. Kirkby Ravens- | 201. Bingley. | 274. Barnsley. |
| worth. | 202. Calverley. | 275. Darfield. |
| 25. Aldborough. | 203. Seacroft. | 278. Brodsworth. |
| 32. Whitby. | 204. Aberford. | 281. Langsell. |
| 33. —— | 215. Pocke Well. | 282. Wortley. |
| 33. Marske. | 218. Bradford. | 283. Wath upon |
| 33. Richmond. | 217. Calverley. | Dearne. |
| 46. —— | 218. Leeds. | 284. Conisborough. |
| 47. Robin Hood's | 219. Kippax. | 287. Low Bradford. |
| Bay. | 231. Halifax. | 288. Ecclesfield. |
| 53. Downholme. | 232. Birstal. | 289. Rotherham. |
| 68. Leybourne. | 233. East Ardsley. | 290. Braithwell. |
| 82. Kidstone. | 234. Castleford. | 293. Hallam Moors. |
| 84. E. Witton. | 246. Huddersfield. | 295. Handsworth. |
| 97. Foxup. | 247. Dewsbury. | 296. Laughton - en - |
| 98. Kirk Gill. | 248. Wakefield. | le-Morthen. |
| 99. Haden Carr. | 249. Pontefract. | 299. —— |
| 100. Lofthouse. | 250. Darrington. | 300. Harthill. |
| 115. Arncliffe. | | |

### MINERAL STATISTICS.

Embracing the produce of Coals, Metallic Ores, and other Minerals. By R. Hunt. From 1853 to 1857, inclusive, 1s. 3d. each. 1858, Part I., 1s. 6d.; Part II., 5s. 1859, 1s. 6d. 1860, 3s. 6d. 1861, 2s.; and Appendix, 1s. 1862, 2s. 6d. 1863, 2s. 6d. 1864, 2s. 1865, 2s. 6d. 1866 to 1881, 2s. each.
(These Statistics are now published by the Home Office, as parts of the *Reports of the Inspectors of Mines*.)

### THE IRON ORES OF GREAT BRITAIN.

Part I. The North and North Midland Counties of England (*Out of print*). Part II. South Staffordshire. Price 1s. Part III. South Wales. Price 1s. 3d. Part IV. The Shropshire Coal-field and North Staffordshire. 1s. 3d.

www.ingramcontent.com/pod-product-compliance
Lightning Source LLC
Chambersburg PA
CBHW022032190326
41519CB00010B/1682